Machines
of the
Ancient World

SCHOOL PUBLISHERS

Orlando   Austin   New York   San Diego   Toronto   London

Visit *The Learning Site!*
www.harcourtschool.com

Introduction

These days, we use many impressive technical gadgets in our everyday lives. Just think of all of the electrical devices you use: televisions, stereos, computers, cellular phones—the list goes on.

You might think that the word *technology* applies only to electronics. However, technology, as it is discussed in science, refers to more than just your DVD player. Technology includes any tools that we use to make life easier. In fact, long before the invention of the computer, humans were clever inventors. Many of the tools we use today are based on devices that were invented thousands of years ago.

Ramps in Ancient Egypt

Have you ever watched people unload their belongings from a moving van? The ramp propped onto the back of a moving van is a simple machine. A ramp is an inclined plane that works by reducing the amount of force needed to move an object. Scientists believe the ancient Egyptians were well aware of the concept of the ramp, and they used it to make their work easier while building the pyramids.

Suppose that you are the manager of a construction team in ancient Egypt. Your team has to lift a boulder weighing 2300 kg (5000 lb) (twice the weight of a car) 15 m (50 ft) into the air. How will you do it? If the team tried to lift the boulder straight up, they would have to apply 115,000 kg (250,000 lb) of force to it. Impossible! But suddenly, you have the idea of building a 60-m (200 ft) slope leading up to the construction site, and the idea for the ramp is born.

Pyramid Work Teams

The builders gathered farmers together when harvest was over to make up construction teams. Bakers came to the makeshift construction towns to bake for the crews. Archaeologists have found graffiti in the towns that suggest that the crews had parties during their off-hour times.

Now, the amount of force needed to move the boulder can be divided by the ramp length, which would mean that the team must apply only 575 kg (1250 lb) of force. Yet this amount of effort is still unrealistic. However, if you put five ropes around each boulder with five workers to pull the ropes, and add five more workers to push the boulder from the other side, no individual has to apply more force than his own body weight, and the boulder can be moved. The ramp allows the workers to push and pull the boulder without having to lift it. As the team manager, you've discovered a simple solution to a difficult problem.

Although we still don't know for certain exactly how the pyramids were built, what you have just read is a plausible description of the process. The Egyptians probably didn't use wheels because they were working in soft sand, and wheels would have gotten stuck. They probably used water or some other type of lubricant on the ramp to make it slippery—just as the rollers on a mover's ramp make it slippery. Both rollers and lubricants reduce friction and make it easier to pull an object across another surface.

Scientists estimate that the pyramid work teams would have been able to place one of the huge boulders every two-and-a-half minutes with the help of ramps. That's amazing proof of the usefulness of these simple machines!

Winches in Ancient China

One of the most basic human needs is that for water. People living in ancient times spent a great deal of time thinking about how to gather and store water.

The need for a daily supply of water limited the number of places in which people could live. Many groups settled near streams, rivers, and lakes. If people wanted to live somewhere that was not close to a body of water, they had to find another way to meet their daily water needs. Eventually, it was discovered that rivers and lakes exist underground. People began digging deep wells in order to use the underground water supply. However, pulling the water out of the wells required the development of another simple machine.

The people of ancient China used a remarkable machine, called a winch, to draw water. Like a ramp, a winch reduces the amount of human effort needed to lift or move a load. A winch is nothing more than a rope wound around a spool with a crank attached. A person turns the crank by hand, and the rope lowers a barrel deep into the well, where it dips into the water. After the barrel fills, the person turns the crank in the opposite direction, and the rope pulls the barrel back up again. People in ancient China used winches to obtain the water they needed for cooking, bathing, and drinking.

Square Well

A winch is a like a one-wheeled wheel and axle. The drum is the axle. The crank is the wheel.

winch

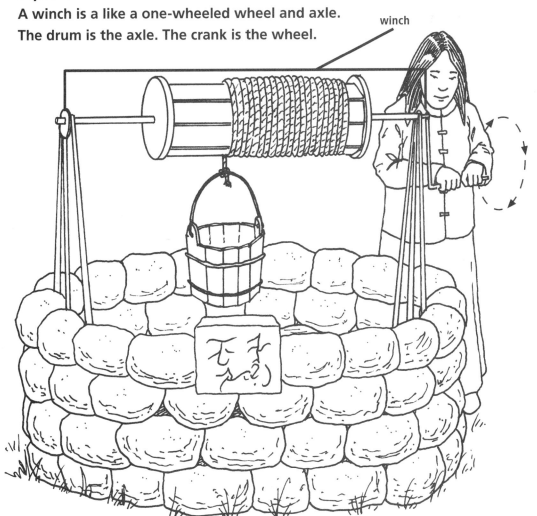

In some areas of China, people still use their ancient winches and wells. The people of Jianshui, a city in southern China, revere their wells as their source of life. Even though the people in Jianshui now have running water, they still get most of their water from the 100 wells throughout the city. People have even named their wells: East Well, Red Well, Triplet Spring, and Dragon Well are just a few. The names of some wells offer a clue about their source. Other names simply state a well's location in the town.

The wells vary in age, from "young" wells that are about 100 years old to wells that are more than 1000 years old. Jianshui prides itself on its wells. The city advertises the wells to attract tourists from all over the world.

The winch also has other uses in the modern world. Mechanics use winches to lift engines from cars. Sailors use winches to raise anchors. This procedure can be quite a feat when you consider that anchors for some large ships, such as aircraft carriers, can weigh as much as 27,000 kg (60,000 lb). Without a winch, it would take the entire crew several hours to raise the anchor, especially on an aircraft carrier. Anchoring ships of that size would not be practical without winches.

Levers: The Product of Ancient Greek Warfare

In many cases, we can't identify the inventors of ancient machines. Just as we don't know who invented the winch, we don't know the name of the Greek soldier who invented the catapult. And yet his invention brought his people many victories in ancient battles.

The Greeks were among the first people to use catapults in warfare. They clearly intended that catapults would be used to destroy an enemy's defenses. To be effective, a catapult had to hurl an object over a long distance with great force.

Catapults are powerful and clever devices, but they're nothing more than simple levers. To see how a catapult works, think of a screwdriver. Have you ever used one to pry open a can of paint? The metal lid is wedged in tight, but if you use a screwdriver as a lever to pry the lid off, the can is easy to open. Using the edge of the paint can as a fulcrum, you place the flat end of the screwdriver under the lip of the lid and apply a downward force on the handle of the screwdriver. Of course, you don't want the lid to fly off and splatter the paint, so you do this whole action carefully.

Soldiers of ancient Greece used similar force with catapults. They used a rock as the fulcrum, placing it off-center so that the lever had one short and one long end. The soldiers probably placed a large boulder in a basket attached to the long end. To spring the boulder into the air toward the enemy, the soldiers would have had to drop a counterweight onto the short end of the lever. You've probably guessed by now that the counterweight for the catapult would have had to be very heavy in order to hurl an enormous weight across a battlefield.

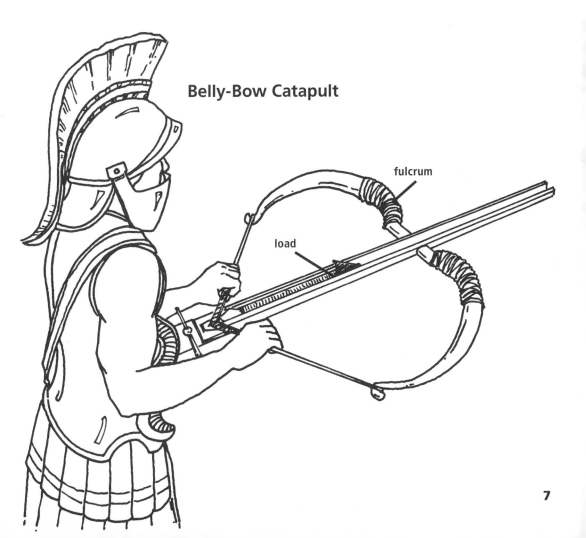

Belly-Bow Catapult

fulcrum

load

The first known catapult, called the belly-bow, was a form of bow and arrow. The two sides of the bow come out from the center rod and arch backward. The string acts the same way as the pouch holding the boulder. But in this case, the arrow is the weapon that is projected toward the enemy.

The belly-bow combined the energy of two torsion springs onto one central rod. It was actually nothing more than two simple levers. The energy applied to the arrow came from the force of two levers working together, so the arrow flew with great power and across a long distance.

The Greeks continued to improve their catapult designs. In addition to simply flexing and bending a rod in order to produce force, they added twisting to the process, which allowed the device to store and release more energy.

They also began to use springs made of large ropes of animal hair. The soldiers pulled back on the handle, and the "hair spring" became twisted, storing up energy. When the spring was released, the catapult shot rocks, arrows, or burning tar onto enemies.

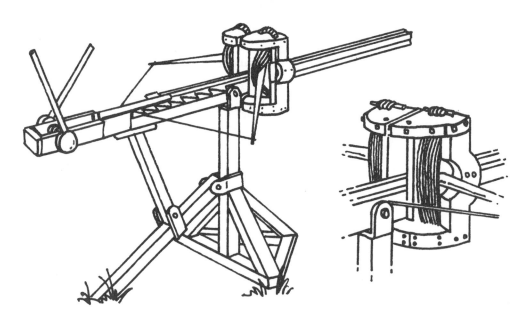

Spring Catapult
Alexander the Great used spring catapults to conquer the world. Later, many Greek city governments made spring catapults by the dozens and used them as a main defense.

The Water Pump

Perhaps the most famous of the early Greek inventors was Archimedes. He lived more than 2000 years ago, and was one of the most famous mathematicians in all of ancient Greece. He was called "the wise one" and "the master."

Archimedes, who was the son of an astronomer, had close ties to the king, Hiero. In fact, Archimedes was often called upon to solve Hiero's most difficult problems.

Archimedes is credited with discovering several laws of physics. Legend has it that, when he discovered one of these laws, he ran through the streets shouting "Eureka!," meaning "I have found it!"

Thinking was Archimedes' main occupation. It is said that he was often so lost in thought that he'd forget to eat. That may seem odd, but in ancient Greece, many wealthy people believed that their most important task was simply to sit and think. Developing theories and discussing scientific law were considered to be the most noble work that a person could do. People believed that tools were meant for the slaves and, because slaves did all of the work, the upper classes thought it was beneath their lofty position to invent tools.

Thinking about mathematics took up most of Archimedes' time. Although he would continually scribble his theories about geometry on any surface he could find—dust on the ground, ashes in a fireplace, and even on his own skin—he didn't think it was important to write down anything about the machines that he invented. Fortunately, because Archimedes was so famous, others wrote about him, so he is credited with inventing many unique devices. One of these was a water pump.

Archimedes' water pump, also known as Archimedes' screw, is a very simple device. It is made by placing a screw inside a cylinder. A crank is placed in one end of the cylinder and attached to the screw inside. The other end of the cylinder is placed in water. When the crank is turned, water travels upward through the cylinder and empties out the top.

Archimedes' Water Pump

Some sources believe that Archimedes developed this water pump while he was studying in Egypt. Other sources claim that he invented it because King Hiero asked him to find a way to drain water out of one of his ships. Egyptian farmers did use this kind of pump to raise water from the Nile River into irrigation ditches. They were astonished at how much water they could bring up with very little labor. Some people use a variation of the water screw to irrigate crops, even today.

The idea behind Archimedes' screw is still used today in some machines, such as boat propellers. A boat propeller works like Archimedes' pump: it moves water. The boat propeller actually causes two motions. First, it pushes water away from the boat, and then the boat moves forward.

You are probably most familiar with screws when they're used as fasteners, but they're also used as parts in many kinds of machines. Screws were used in an early Greek invention called a screw press. Greeks made a screw press from two flat pieces of wood with a screw in each of the four corners. When the screws turned, they produced a force that pressed the two pieces of wood together. When fruits or vegetables were placed between the pieces of wood, turning the screws squeezed out juice or oil. In this way, the Greeks could get grape juice from grapes and get olive oil from olives. Screw presses are still used today.

Pulleys

According to a popular legend, Archimedes once bragged to Hiero, "Give me a place to stand, and singlehanded I can move the Earth." If this legend is true, Archimedes may have been talking about the math equations he discovered that explain how to balance weights with pulleys.

Hiero was curious and asked Archimedes for proof of the claim that he could "move the Earth." According to one writer, Archimedes chose a cargo ship from the royal fleet and gave orders for it to be loaded with passengers and goods. The story goes on to say that Archimedes sat down on the dock, holding onto a rope threaded through a complex system of pulleys, and pulled the large, heavy ship toward him without seeming to use any real force or effort.

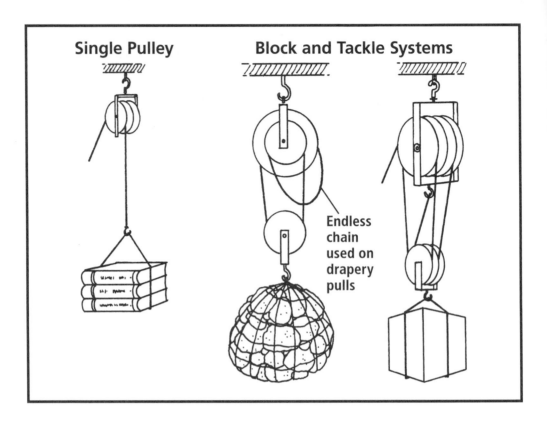

Single Pulley

Block and Tackle Systems

Endless chain used on drapery pulls

No one is sure exactly what kind of machine Archimedes used to perform this feat, but some guess that it was a block-and-tackle system, which is a tool made from many pulleys and hooks. A single pulley works by changing the direction of the force applied; it does not increase the force used to move an object. But a block-and-tackle system can double the force applied to an object, making it much easier to move a large object.

Many scholars think that this story about Archimedes is exaggerated. They believe that Archimedes probably demonstrated his theory with a much smaller object, perhaps a small fishing boat. But no one doubts that Archimedes used a compound system of pulleys to carry out the task.

We still use pulleys in many everyday activities. If you have drapes or large, thick curtains in your living room, you may use a compound system of pulleys when you pull on the cord that opens and closes them. This is the same principle that Archimedes used to drag the ship. Although the drapes may be heavy, the system of cords and pulleys allows them to slide open easily.

Planetarium and War Machines

Archimedes was also responsible for another exciting invention—the first planetarium. Archimedes' planetarium was not a big, dark building that uses a projector to show the stars on the ceiling, such as the kind of planetarium that you may have visited. His was a complicated maze of gears and spheres that imitated the movement of the stars, planets, and Earth's moon in their orbits. A large transparent sphere, possibly made of glass, enclosed the entire mechanism so that people could see the smaller objects moving inside.

Archimedes thought of his planetarium as a concrete demonstration of the math equations he had figured out for the orbits of the planets. Perhaps that is why he thought this invention was more important than the others—he even left a written record about it.

Planetarium Machines

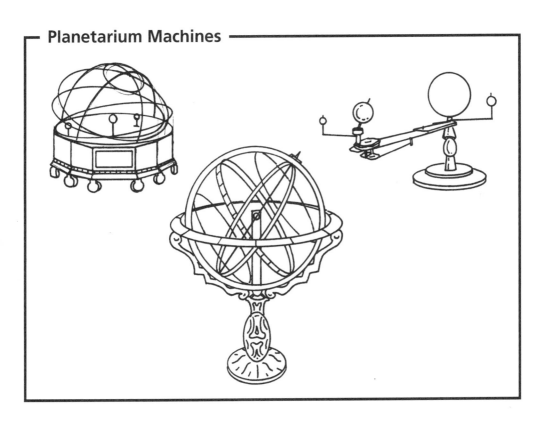

Archimedes' planetarium was a mathematical feat so incredible that no one, even today, has been able to figure out how he accomplished it. News of the planetarium traveled outside Greece and all the way to the Greeks' enemies, the Romans.

During this time in history, the Greeks and the Romans had begun battling each other. Archimedes was responsible for developing some of the war machines that helped the Greeks succeed in battle.

Archimedes used a spring-lever system to devise a very powerful version of the catapult. The Greeks used this weapon effectively to defeat their enemies in battle. He also invented an iron-clawed crane that could lift ships from the water. (It is likely that this crane also operated on some sort of pulley system.) When the crane's claw dropped a ship, the ship was smashed to pieces in the water or against cliffs and rocks. Some say that

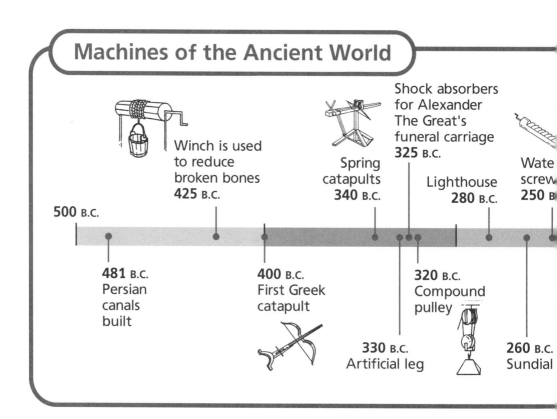

Machines of the Ancient World

Winch is used to reduce broken bones
425 B.C.

Spring catapults
340 B.C.

Shock absorbers for Alexander The Great's funeral carriage
325 B.C.

Lighthouse
280 B.C.

Water screw
250 B.

500 B.C.

481 B.C.
Persian canals built

400 B.C.
First Greek catapult

330 B.C.
Artificial leg

320 B.C.
Compound pulley

260 B.C.
Sundial

the Roman soldiers grew so afraid of Archimedes' powerful inventions that, if they saw so much as a rope appear over the top of a wall, they would turn and run, yelling, "Look, Archimedes is aiming one of his machines at us!" Archimedes' inventions helped the Greeks hold off their Roman attackers for many years.

Eventually, Archimedes lost his own life in battle. One Roman general devised a plan to capture Archimedes and force him to join the Roman army, thinking that the great thinker might introduce them to some of his "war machines" and help the Romans to win battles. Unfortunately, however, a Roman soldier came upon Archimedes while the inventor was, as usual, lost in thought. One legend says that Archimedes was drawing geometric figures in the dust. According to this story, Archimedes said to the soldier, "Don't disturb my circles!" The soldier didn't recognize Archimedes, and killed him immediately.

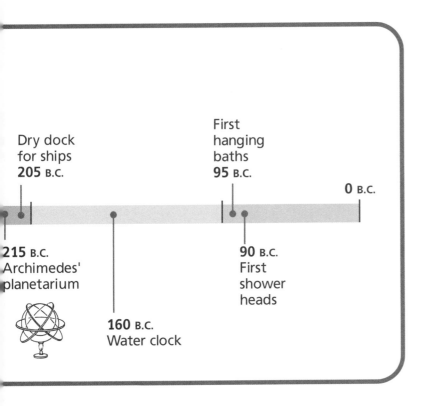

Dry dock
for ships
205 B.C.

First
hanging
baths
95 B.C.

0 B.C.

215 B.C.
Archimedes'
planetarium

90 B.C.
First
shower
heads

160 B.C.
Water clock

When the Romans learned of the tragedy, they were so sorry that they gave Archimedes a burial ceremony and engraved his favorite math equation on his headstone. The planetarium was taken back to Rome as a war treasure. No one today knows where the planetarium or the grave site is, but no one has forgotten Archimedes' great machines.

Ideas and Inventions

Physical laws that govern the action of machines have a direct effect on how we live our lives. As you have read, many of the gadgets that we use in everyday life came from ancient ideas. Humans have always worked to create new tools that make life easier. Curiosity, creativity, and practical tinkering have allowed humans to survive and succeed in their environment.

As human history has progressed, scientists have learned that they must write down their new ideas in order to save them, to teach others, and to build on them for later scientific discoveries.